A souvenir pictorial record of th
airshow held at Omaka Aerodro

Editor: Allan Udy

Photographers
Alex Mitchell   Philip Merry
Geoff Sloan   Phil Teague
Jim Tannock   Karen Mitchell
Craig Justo   Paul Charal
Paul Maggs   Lawrence Ackett
Chris Guy   D.L.A. Turner
M. Hodgkinson   Jane Orphan
G. Hadfield/Langwoods Photo Centre
Rob Duff/Topshots Photolab

Additional Contributors
Allan Baker   Joanna Carson
Dave Lochead   Stefan Schmoll

Design/Production/Publishing

Golden Micro Solutions Ltd,
Box 590, Blenheim, NZ.
http://www.golden.co.nz

Printing
Wyatt & Wilson Print Ltd,
Christchurch, New Zealand
http://www.wyatt-wilson.co.nz

Copyright © 2005 Golden Micro Solutions who assert their moral rights in this work. All images remain copyright their respected photographers. This Souvenir Edition was independently produced, with approval from, and in consultation with Classic Fighters Marlborough.

This publication is copyright, and other than for the purpose of fair reviewing may not be reproduced or transmitted in any form, electronic or mechanical, photocopying, or recording, without the written permission of the publishers.

ISBN 0-473-10288-9

www.classicfighters.co.nz

Cover Artwork: Spike Wademan,
Box 9, Queenstown, NZ. Ph: 0274967076

Back Cover Photograph: Gavin Conroy, T.E. Shaw (Lawrence) Brough Superior vs Bristol Fighter F2b race reenactment.

It Seemed Like A Good Idea At The Time...

Somewhere between Easter 2003 and Easter 2005, someone made the suggestion to me that Classic Fighters should produce a souvenir book of photos from the 2005 show. It seemed like a good idea at the time—get it all ready before the show, minus the captions and photos, and then after Easter get it finished, printed and out the door within six weeks or so. As we all know, the best laid plans don't always work out. By the time January 2005 rolled around, I was working pretty solidly on general airshow organisation issues, as were many of my fellow committee members, and the idea of trying to cram a few more hours of work into the day didn't really appeal.

So here we are some five months after the show, and I'm preparing the final parts of the 'book' before I send it away to the printers. The publishing of this book is now a private venture, rather than a direct Classic Fighters project, and it's taken longer than planned, and hoped. However, I think the end result is going to be worth the wait. Just like Classic Fighters, and the Aviation Heritage Centre....

The AHC and Classic Fighters started life as crazy ideas thrown around amongst a few friends. We've been through three airshows, and the initial building stages of the AHC, and at times it's felt like we were the 600 riding into the Valley of Death—guns to the left of us, guns to the right, and guns straight ahead. Some of those ideas turned out to be much harder to implement, or even crazier, than we first thought—but they all seemed like good ideas at the time.

I think what sets Classic Fighters, the AHC, and Omaka apart from the rest, is that not only do we have lots of good ideas (along with a few not so good ones now and then), but we also have a community with the drive, commitment and foresight to make those ideas a reality. Sometimes it may take a little longer than anticipated, but then all good things take time. I'm proud to be a part of that community turning some of these ideas into reality.

In this Souvenir Edition I've tried to cover as much of this year's show as possible, and to provide a good retrospective feel for the event. I've probably missed including a few critical photos, or not mentioned someone I should have named. Please accept my apologies in advance if that's the case, as it's certainly not intentional. So many people from this community, and from our military, aviation and warbird friends around New Zealand, as well as our sponsors, have helped make this event the great success it is, that it's an almost impossible task to cover and thank everyone involved. I'll do it now—thanks!

The best thing about getting to this stage is that now I've finished the book; I can start work on the movie. Seems like a great idea!

Allan Udy - Editor

Classic Fighters Marlborough 2005   1

The Marlborough Region

- Havelock, Pelorus and Kenepuru Sounds
- Picton and Queen Charlotte Sounds
- Blenheim, Renwick and Surrounding Region
- Awatere and Pacific Coastline

Facts about Marlborough

GETTING TO MARLBOROUGH

- The main airport for the province is located just 10 minutes west of Blenheim in Woodbourne.
- Picton Airport is located at Koromiko on SH1, approx 5 km from Picton
- Omaka Airfield on the southern outskirts of Blenheim. Marlborough Aero Club and recreational flying.
- Picton is the South Island port for the Cook Strait ferries

DRIVING TIMES IN AND AROUND MARLBOROUGH

JOURNEY	KILOMETRES	DRIVING TIME
Kaikoura to Blenheim	129 km	1.5 hours
Nelson to Blenheim	116 km	1.5 hours
Christchurch to Blenheim	308 km	4 hours
Nelson to Picton	144 km	2 hours
Christchurch to Kaikoura	178 km	2 hours
Blenheim to Picton	30 km	30 minutes
Blenheim to Seddon	30 km	30 minutes
Blenheim to Wairau Valley	30 km	30 minutes
Blenheim to Havelock	40 km	35 minutes

For more information about local attractions, transport and accommodation

VIN ACCREDITED INFORMATION & TRAVEL CENTRES

Blenheim — Blenheim Railway Station SH1, Blenheim.
T: 03 577 8080, F: 03 577 8079,
E: mvic@destinationmarlborough.com

Picton — Foreshore. T: 03 520 3113, F: 03 573 5021.
E: pvic@destinationmarlborough.com

Kaikoura Information Centre — Westend, Kaikoura.
T: 03 319 5641 F: 03 319 6819
E: info@kaikoura.co.nz

REGIONAL STATISTICS

Marlborough has a moderate climate; typically long dry summers and crisp, clear winters.
Summer (Dec-Feb) 24°C
Autumn (Mar - May) 13°C
Winter (Jun - Aug) 10°C
Spring (Sep - Nov) 14°C

Marlborough region population 41,700
Blenheim township population 27,900
Picton / Waikawa population 3,990
Sunshine hours 2469
Average annual rainfall 673mm

If you are dialling a New Zealand telephone or facsimile number from outside New Zealand the country code is **64**

http://www.destinationmarlborough.com

Chairman's Message

Welcome to the 2005 Classic Fighters Marlborough Souvenir publication.

I had to write a piece earlier this year in anticipation of the show. At that time I thought it important to write briefly about the gifts that this committee had been so fortunate to work with as they prepared for the 2005 show. With a few changes in tense, and the benefit of hindsight, I repeat largely what I said then.

Omaka is a bit a shrine for me; I grew up in Marlborough but lived away from here for more than 25 years. Often, and particularly while I was overseas, my thoughts turned to this, my first home. I frequently thought about this special place where New Zealand's first ever air pageant was held in 1930. I love the hills to the south, especially on a late summer evening when the shadows paint the contrasts of the contours so vividly. My father, who worked for Safe Air from 1951 until he retired in 1976, worked on the very Bristol Freighter by the Marlborough Aero Club clubrooms; I'm tied to this place. I love its natural amphitheatre. Thank you to the Marlborough Aero Club for granting us this gift of your aerodrome for the weekend.

I want to acknowledge the gifts we have received from the aviation community throughout New Zealand. As with so many Kiwi endeavours, we seem to be able to punch well above our weight. From our tiny population in New Zealand, and to a large extent from Marlborough, there have emerged enough enthusiasts; aircraft restorers, owners, and pilots to enable us to borrow more than 80 aircraft for this event. We worked with those gifts to ensure that we brought those aircraft to Omaka in the most creative way we could. We wanted to show aircraft in the context of action and stories not found at many airshows anywhere in the world. Thank you to all those restorers, owners, and pilots.

We were also fortunate to be given the priceless gift of community backing for this event. From an amazingly supportive community we have received materials, enthusiasm, and commitment, reinforcing the motivation of more than four hundred volunteers. It was from that gift that the committee's enthusiasm was sustained. Thanks Marlborough.

One of our objectives was to raise money to help fund the Aviation

Karen Mitchell

Heritage Centre at Omaka which will be a lasting gift to future generations. Unfortunately, we had challenging weather throughout a lot of the weekend and numbers were down on what I thought we deserved. But it could have been worse and the photographers had dark and dramatic clouds in which to frame their aircraft shots. There were no incidents and everyone that I could see went away happy. Not a bad result overall!❖

Allan Baker, Chairman,
Classic Fighters Charitable Trust

A Dedicated Team

An event of the size and scope of Classic Fighters requires a significant amount of manpower, a huge amount of organisation, and a healthy dose of political machinations along the way! As mentioned above by Allan Baker, this year the event received the generous help from over 400 community volunteers: people manning the gates in the rain for a morning, through to suppliers of services required on-site, to those more involved with the actual display aspects of the show. Without these people the show wouldn't have been possible, and for that we thank them all. Yet we must also remember that without the dedication of the organising committee team at the helm the show would not have happened either. Take a look at the folks in the photo below—between them they have spent well over a thousand hours of volunteer time in the leadup to Easter 2005, to bring you the show that we review here in this publication. If you have the opportunity, do thank them, and let them know that their mammoth effort was worthwhile, and well appreciated.❖

The Classic Fighters 2005 organising committee: Jay McIntyre, Rob Lissaman, Mike Nichols, Dave Lochead, Tony Clarry, Shirley Clarry, Allan Baker *(Chairman)*, Allan Udy, Brian Greenall, Willy Sage, Lester Hope. Absent: Tim Sullivan, Steve Petersen   Photo: Erin Boyd

Aviation Heritage Centre - Vision To Reality

Artists Impression

While the aim of Classic Fighters has been to provide our guests with a world-class airshow and entertainment event, the primary goal has always been to raise money to help build and develop the Aviation Heritage Centre at Omaka Aerodrome.

Initially discussed in the mid-1990's by members of the (then) fledgling Marlborough Warbirds Association, the development of the centre has been a long, but ultimately rewarding experience for those involved. Visitors to Omaka Aerodrome at Easter 2005 had the opportunity to witness that the idea from the mid-90's, having evolved to an artists impression (above) in 2003, has now almost turned to reality.

Construction of the first two 30 x 50m hangars began in late 2004, and by Easter 2005 the work was nearing completion. When the Centre is opened to the public in late 2005/early 2006, these two hangars will house the largest collection of full size World War One aircraft in the world.

The initial objective was to house the WW1 aircraft in one hangar, and use the second for many of the other notable warbird and historic aircraft permanently based at Omaka. However, the continued growth of the number of Great War aircraft available for display has provided the New Zealand Aviation Museum Trust with a wonderful opportunity to create a unique, world class facility right from day one.

Efforts are even now underway to continue the fund raising programme so that construction of the third hangar can begin as soon as possible. It is hoped that within a short space of time that work will be completed, providing display and storage space for some of the other Omaka aircraft.

The feature that will ensure the AHC stands out from other aviation museums around New Zealand, and indeed worldwide, is that the centre will be a 'living' museum. While some of the exhibits will be necessarily static, a significant portion of the aircraft to be displayed in the museum hangars are, and will remain, fully airworthy. It is fully expected that in years to come, the biennial Classic Fighters airshow will not be your only opportunity to see these wonderful aircraft in the air.❖

Jane Orphan

Top: Artists impression of the Aviation Heritage Centre.
Above: The view 'inside' one of the almost completed hangars in January 2005 shows the size of the building.
Below: The AHC as seen from the main field at Easter 2005. Access was limited due to continuing construction on site, but the sight bodes well for the future of historic aviation at Omaka Aerodrome.

Geoff Sloan

Classic Fighters—The Early Years

Three factors combined to create Classic Fighters Marlborough.

The first was that many of New Zealand's warbird pilots had chosen beautiful Omaka Aerodrome as the ideal Easter getaway destination in those years they weren't performing at the famed Warbirds Over Wanaka show. Centrally located, handy to some of the world's best wineries and historically enjoying some of the best Easter weather in the country, these impromptu gatherings grew in popularity. With them grew the crowds of onlookers travelling from as far away as Nelson on the assumption that the warbirds would be in town.

At the same time, a little known film director by the name of Peter Jackson cranked up his deep interest in WW1 aviation history, by asking vintage aircraft builder and pilot Stuart Tantrum to get a Sopwith Camel and Fokker Triplane (among other unimaginable treasures), into the air.

The third factor completed the jigsaw. Local resident Graham Orphan, editor of aviation magazine Classic Wings, already had a history of making things happen in the Australian aviation scene. He might not have been the only person in Marlborough who saw the obvious, that the elements for an airshow were already in place, but it was he and his wife Jane, plus a small group of warbird enthusiast friends, who decided the first event would not be just any old airshow.

It wasn't long before a group of volunteers were forming impossibly big visions. With no money and no track record, their goal was to create an international event of note, right off the blocks.

Marlborough, home to the Royal New Zealand Air Force and aviation engineering firm Safe Air Ltd, is a melting pot of creative and engineering talent. Non-aviators with specialist skills were dragged in, many developing a lasting interest in the scene. It was a great basis for such a mammoth task, but it wasn't enough. Slowly, with financial and in-kind support, the entire community swung in behind the event. They did so with nothing but the infectious enthusiasm and questionable confidence of a bunch of mad aviators to back up their investment.

It was hard, hard work, and members of the core team, which didn't grow any bigger as the ideas took on ever increasing scale, try not to think back on it. But the effort seemed worth it when, at Easter 2001, the inaugural Classic Fighters Marlborough airshow took to the air. There over Omaka were the glorious sounds of a Sopwith Camel, a Fokker Triplane and a Hawker Hurricane. More impressive than the fact that these three special aircraft were New Zealand's own was the fact this first-time show featured two significant imports—a WW1 Bristol Fighter replica from the USA and the crowd-favourite WW2 CAC Boomerang fighter from Australia.

The two-day show attracted 25,000 people, and they had a ball. Part relaxed family-minded country carnival, part colourful military re-enactment, it was nevertheless every bit an international airshow. It was unique yet right up there in quality, and with over 40 classic aircraft on the field, in quantity too. The show had a number of New Zealand firsts. It had scenes the crowds had never before witnessed, and it had a spectacular scenic backdrop for all this.

By 2003, the number of rare aircraft had trebled. Instead of one Triplane to battle the Bristol Fighter (which never went home) and Sopwith Camel over the vineyards, there was a staggering five. New Zealand firsts became world firsts – at least since the First World War! Everything (but the core organising group) was ramped up in size—from the props to the ground theatre, special effects and the aircraft scenarios themselves. If anything, the work was harder than before, but the general consensus was that the result was bigger and better.

Which brings us to Classic Fighters Marlborough 2005. Bigger and better yet? You bet. By now, despite the impressive advances in aviation which have occurred in this region since Classic Fighters 2001, the volunteers were asking themselves why they were still putting themselves through such punishment? I think you will agree that the answer lies on these pages. ❖

Joanna Carson - '01/'03 Committee

Top: The WW2-era CA-19 Boomerang was one of the 2001 stars. **Left:** In 2003 one of the Harvards was repainted as an Italian Maachi for the WW2 North African scenario. **Above:** Southward Car Museum brought their Mercedes 770k as a special guest vehicle in 2003. **Below:** 2003—the first time since 1918 that five Fokker triplanes flew together.

Classic Fighters Marlborough 2005

Classic Fighters 2005 - A Review

The best way to describe Classic Fighters 2005 is to paint a picture of the day before the show. Like all shows the field is a hive of activity. Tents are going up, the big screen is being unpacked from its container and installed at the front of the crowd line (later in the day WW2 fighters will disappear behind this screen, so close and low can they fly in this unique setting). Final aircraft are arriving in dribs and drabs, or are taking off on practice sorties. Harassed organisers meet in worried clusters to discuss the latest (bad) weather reports and to compare notes on which aircraft have been seen and which are still to arrive. Why are they not here? What if they can't make it?

All this is normal, but at most airshows you'll not find a team from Weta Workshops (the movie special effects company) laying out the very realistic bones of a WW1 and WW2 battleground. Not at every airshow will you find a local engineer who has taken leave from his job to build a 20m tall replica of the Eiffel Tower out of scaffolding. You might struggle to find a working windmill, complete with mood lighting and rotating blades, at the entrance to your gold pass marquee—whipped up by someone who thought it would just complete the Moulin Rouge theme. Or you might not be used to seeing a petanque court, a German Wurzburg radar array (built in someone's back yard), or even the Arc de Triomphe!

But they were all there, and they combined to add flair and a carnival atmosphere to the French-themed extravaganza. The aircraft types available, and Omaka's setting as a multi-vector grass field surrounded by vineyards and a backdrop of bare hills, makes it difficult to go past France as a theme. Classic Fighters 2001 and 2003 based the WW1 scenarios in that setting, but for WW2 moved on to the Pacific (in 2001), and North Africa (in 2003). But there are still many French stories from both wars yet to be told, and it wasn't difficult to find an antipodean involvement for 2005.

This is why the battle of Le Quesnoy was reenacted in front of a crowd of around 20,000. Not only did they experience the incessant clatter of small arms fire, smell the smoke and imagine the relief that would have ensued when New Zealand troops liberated the village, but through the commentary they learned the unknown history of a battle in which Kiwi forces did themselves proud.

How well it was reenacted. Many found it difficult to tear themselves away from the ground theatre drama just metres away, and almost forgot to look up to see nine or ten WW1 aircraft tossing and tumbling in the sky above. Upstairs, where the five Fokker Triplanes, the Bristol Fighter, Sopwith Camel, Halberstadt, Pfalz D.III, DH-2 and DH-5 were staking their claim, the story continued. When one side was beaten, its aircraft peeled away to hide at low level behind trees to the south of the field, trailing the ubiqitous smoke of the injured.

A second special story was played out with great aplomb later in the day. But its true significance can only be appreciated by painting another picture, this time of the pilot's briefing on Saturday. Triplane pilot John 'von Richthofen' Lanham, later to portray the last flight of the German ace, is handed a box. The contents are entrusted to him for the duration of the weekend, and are revealed into an awed silence. In the box is the very scarf taken from the Red Baron as he lay mortally wounded in his aircraft after making his last ever landing. For pilots quite used to the thrill of the display, moments like this are as real as it gets.

The re-enactment itself, the final moments of which were played out in front of the gold pass stand, was much enjoyed, both by the crowd and the actors. So popular has the ground theatre become that the big screen, a new initiative this year, was a hit with those not early enough to stake a claim on the fence line. They didn't have to miss a thing, and a relaxed picnic behind the crowd line proved for many a far less stressful way to enjoy a great family experience.

The programme was so full there was not even time for a lunch break this year. At noon the RNZAF displayed its parachutists, formation aerobatic team, helicopters, maritime patrol and VIP aircraft. On a field this intimate, the low-level formation passes by an Orion and a Boeing 757 are unquestionably the highlight for many. During the day a number of exciting aerobatic routines by the Edge 540, Giles 202 and Pitts Special also thrilled the crowd. For sheer skill and thrill, the agricultural spraying demonstration must be firmly put in this category as well.

Breathless as many visitors were by this time, there was just as much to come. Famous Kiwi aviator Jean Batten flew in to join the celebrations in her Percival Proctor, as did some commuters of the golden age in some classic early airliners. WW2 was not forgotten either, even though the size of the show's WW1 contingent could easily leave the

heavy metal in the shade. For WW2 Classic Fighters constantly reinvents itself, by adding removable paint, wonderful ground-based support and compelling scenarios to tell real stories. So the odd aircraft up there might not be quite what it's pretending to be, but that just adds to the fun. There's nothing fake about the stars of this show, though. We saw the best. Mustang, Spitfire, Corsair and Kittyhawk flown by some of the world's most respected airshow pilots, including Ray Hanna and Keith Skilling. But here they become something more—stars in a most realistic drama. That, along with the curved crowd line, draws the very best, and easily the closest, displays of aviation showmanship you will see anywhere in the world.

This year a decision was made to focus on those aspects of Classic Fighters which are unique, so most of NZ's jet warbirds were not in attendance. However their end was ably held up by a spirited performance by Pat Donovan's Blenheim-based Soko Galeb—a new sight in Kiwi skies.

Nothing is perfect. There were a few not-so-welcome visitors such as a cold wind on Friday, some showers on Saturday and strangely, a swarm of annoying flying insects which even the locals had never seen before. Not quite the settled Autumn weather Omaka expects at Easter. But we'll get it in 2007, and true to form the plans are already growing. The hot tip? You ain't seen nothin' yet!❖

Joanna Carson - '01/'03 Committee

Construction Volunteers

One of the things that has always helped Classic Fighters stand out from other events is the amount of time and effort that has been put into developing and building the various props for the shows. Committee member Lester Hope (a Safe Air avionics engineer), has been involved with 'prop development' for all three shows, and for the last two he's managed to assemble a large and willing team of volunteers to take on the challenge of making Classic Fighters a unique event.

Back in 2001, the props team built a WW1 tank and amazing French Chateau, amongst other items. For our *North Africa* themed show in 2003 the team built a 9m high pyramid, an Egyptian tomb (Gold Pass marquee), and a full size Junkers Ju87 Stuka dive bomber, plus many smaller props.

When it was decided that the theme for the 2005 show would be 'France', the race was on to see how many great ideas for props the team could come up with, and how many could be realistically constructed.

It should be no surprise that when Lester first mooted the idea of building a replica of the Eiffel Tower, in excess of 20m high, not too many people took him seriously. However Lester's not the sort of chap to let that get in his way, and so he set about planning how to erect a replica tower constructed entirely from scaffolding materials. Visitors to the show at Easter 2005 will no doubt remember the results of his tenacity and vision—'our' Eiffel Tower certainly helped to add to the French feel and flavour of the show.

The props team also excelled themselves with numerous other items including the Moulin Rouge, some superb sentry boxes, a German Nebelwerfer rocket launcher, a walled French chateau, the French town square, the magnificent German Wurzburg radar array, and even our own version of the Arc De Triomphe!

The photos on this page and the next celebrate the achievements of the props team, and highlight some of the team members. In addition, thanks must be given to the following volunteers who also contributed to this years show: Chris Richards, Bruce Laurenson, John Mahon, John Evans, Ian McKercher, Tim & Jess Hope, and Graham Robson.❖

Top: Lester Hope and Darryn Stewart working on the construction of the Eiffel Tower. **Above:** A passing helper, gives Lindsay Jefferis (right) a helping hand to erect the Wurzburg radar array. **Left:** Lester Hope putting the final coat of paint onto the top unit of the Eiffel Tower before it's fitted to the completed structure.

Top Left: The Moulin Rouge, as built by Ron & Faye Manning. **Top Right**: Ross Erickson preparing the stunningly detailed panels for the Arc De Triomphe, built along with wife Trish Erickson. **Centre Left**: Lester Hope guiding and attaching the top of the Eiffel Tower to the base. **Centre Right**: Part of the French town square with appropriate facades were built by Ross Erickson and the Marlborough Boys College art students. **Bottom Left**: One of Alan & Jan Graham's superb sentry boxes. **Bottom Right**: The Wurzbug radar array (built by Lindsay Jefferis and Lester Hope, with help from Roger Lauder) and Dave Lochead's German Nebelwerfer rocket launcher.

A European Perspective

Stefan Schmoll, an aviation writer from Germany visits Omaka.

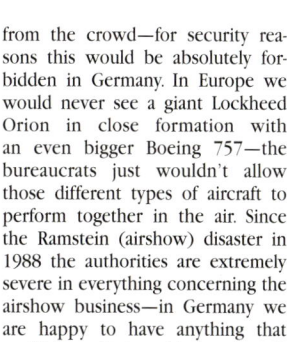

As a regular reader of *Classic Wings* magazine I already had some knowledge of the 2001 and 2003 Classic Fighters airshows. In 2004 I made a trip to New Zealand with my wife and had the opportunity to attend the Taupo Air Show. We were both absolutely fascinated by the show—all those aircraft performing very close in front of the crowd. When I stopped by the Classic Fighters tent to purchase a copy of the Classic Fighters 2003 video, and mentioned my amazement to Allan Udy, his reply was short: Come next year to Omaka and then you will see a **really** good show.... At the time, I never dreamt of being back in New Zealand just one year later!

I arrive in Blenheim, two days before the show, with huge expectations. While I think that Second World War warbirds are great, I was especially keen to see the five World War 1 Fokker Dr.1 triplanes. I've had the privilege of seeing the Breitling Fighters Display four times, have seen many fantastic shows at La Ferté Alais (Salis Collection) in France, and have seen the Flying Legends (Duxford) and the Shuttleworth Collection (Old Warden Aerodrome) in the UK. All of this means when it comes to airshows, it is quite hard to please me... Now I am here with my media accreditation so that I can write several articles about Classic Fighters for the web site www.airventure.de, and for the German aircraft magazine *Flugzeug Classic*. In Europe, we also have spectacular aircraft collections and airshows, but this panorama of the aerodrome with the runway between vineyards and hills in the background is very special.

In the early morning hours the classic biplanes are rolled out of the hangars. Here they are: my beloved Fokker Squadron basking in the sun together with many other aircraft of that era. I see many German and Allied ground troops talking together and suddenly feel like I'm in Paris as the Eiffel Tower and the Moulin Rouge catch my eye. Wow! What an effort for an airshow. This is going to be a great weekend for me. More and more visiting aircraft from other parts of the country arrive, and I'm particularly impressed with the de Havilland biplanes. In 2004 I saw the DH-89 standing in the sun at Gore, the DH-83 with folded wings inside its hangar and the DH-90 at Wanaka airport. Now I have the chance to see all three together in the air, very close to me in front of this wonderful panorama with a perfect sun for photographers (well, it is Friday).

On Saturday and Sunday the stars of the show in my opinion are the Douglas Dakota and Catalina which I saw at Taupo in 2004, and interestingly the Curtiss P-40 and FG-1D Corsair which I already know from the Breitling Fighters team. I'm amazed that the warbirds and all the other aircraft are such a short distance from the crowd—for security reasons this would be absolutely forbidden in Germany. In Europe we would never see a giant Lockheed Orion in close formation with an even bigger Boeing 757—the bureaucrats just wouldn't allow those different types of aircraft to perform together in the air. Since the Ramstein (airshow) disaster in 1988 the authorities are extremely severe in everything concerning the airshow business—in Germany we are happy to have anything that could be called an 'airshow' at all.

I am really very impressed that there are some aircraft with new paint schemes put on especially for the show. The huge Catalina is a wonderful sight in its French Navy livery, and many great photographs of the aircraft and its display have been subsequently published in many European aviation magazines. In addition to all the spectacular things in the air, extraordinary ground theatre scenarios are taking place, and not to be forgotten is the really wonderful relaxed audience. Sometimes I feel more like I'm enjoying a garden party than being at an airshow.

If I think about the things Classic Fighters has in the pipeline for 2007 I'm really excited, especially by all those new classic WWI biplanes—I am really looking forward for the next show even if I only get to see it on video. Now, if you want to really impress many more people from Europe, I have a really good idea: Just bring the five Fokkers to Europe for our airshow season!! ❖

Top: The five German Fokker Dr. 'triplanes' swoop in to attack the airfield on a cold grey March morning.

Left: Visitors to the show relax in front of the French Arc de Triomphe.

Sopwith Camel

One of the crown jewels of WW1 aviation at Omaka is the rotary-engined Sopwith Camel owned by film director Peter Jackson. The aircraft made its debut at Classic Fighters 2001, and along with its American pilot Gene de Marco, has been a crowd favourite ever since.

The aircraft wears the colour scheme of Captain Clive Collett, a Marlborough-born ace of the First World War. A highly respected and competent airman, Collett was killed in a flying accident in 1917, but not before he became engaged to be married. This year we were pleased to be able to welcome his London-born grand-daughter, Mandy Perry, to Omaka. The opportunity to see the replica of her grandfather's Camel flying was too much of a drawcard, and tempted Mandy to make her second trip to New Zealand from Australia within 12 months.

Jim Tannock

Philip Merry

Gavin Conroy

Top Right: The Camel turns sharply to follow the Baron. **Above**: Side-slipping in to land at the eastern end of the field. **Left**: von Richthofen appears over the field before the Camel can take off. **Below**: A low and fast pass for the benefit of the crowd. **Facing Page**: The Camel chases the Baron. Photo Jim Tannock.

Jim Tannock

Fokker Dr.1 'Triplane'

Omaka remains the only place in the world where you can see five Fokker Dr.1 Triplanes in the air at the same time. This extraordinary event first occured in 2003, and now also in 2005. It's believed that this many triplanes have not been airborne simultaneously since 1918. All Dr.1s flying in the world today are replica aircraft as the sole remaining original was destroyed by the Allied bombing of Berlin during WW2.

Jim Tannock

Geoff Sloan

Geoff Sloan

Alex Mitchell

Gavin Conroy

Jim Tannock

Top Right: Five Fokkers equals 15 wings—from left to right, Richthofen, Kempf, Kirschstein, Müller and Greim **Centre Right**: The aircraft of Hans Müller, Fritz Kempf, and Manfred Richthofen take to the air. **Top Left**: Fritz Kempf's aircraft swoops in low, plainly displaying his name painted on the top wing. **Centre Left**: Richthofen, Kirschstein and Greim fly by in close formation. **Above**: Robert von Greim's aircraft on final approach. **Left**: Fritz Kempf's aircraft is piloted by Paul Hughan. **Facing Page**: The Fokker Staffel sweeps in from the west to attack the airfield. Photo Alex Mitchell.

Bristol Fighter F2.b

Along with the Sopwith Camel and a single Fokker Dr.1, this aircraft was one of the original three WW1 stars of Classic Fighters 2001. Built by Ed Storo in the USA, the replica aircraft originally wore a silver, post-war colour scheme, which survived on the aircraft until it was repainted in 2005.

The aircraft represented by the new scheme was transferred from No. 20 to No. 16 Squadron RFC in mid February 1918. While with 16 Squadron, the aircraft retained its previous 20 Squadron markings, and was flown by Captain C. Jones. The aircraft was also flown by Major C.F.A. Portal who went on to become the Chief Of Air Staff from 1940 to 1946.

Philip Merry

Geoff Sloan

Gavin Conroy

Top Right: The Halberstadt D.IV desperately tries to evade the Bristol during a dogfight! **Top Left**: Pilot Tim Sullivan, and visiting guest 'gunner' Fred Murrin, prepare to take to the air. **Left**: The Bif makes a low and fast pass during the practice races prior to Easter. **Below Left**: A low banking pass enables the gunner to 'pepper' the crowd with his Lewis gun, surprising everyone. **Below**: A beautiful machine on a miserable day. **Facing Page**: The Brisfit roars overhead after a practice race with the Brough Superior motocycle. Photo Gavin Conroy.

Paul Charal

Philip Merry

Halberstadt D.IV

This replica is the only example of this aircraft type in the world today, and given that only three originals were built by the Germans during World War One, it's not surprising that none of them has survived.

The D.IV first arrived at Omaka in 2002, and it did make a brief appearance at Classic Fighters 2003 when it flew on the Friday morning session. Unfortunately operational issues intervened, and the aircraft remained a static display aircraft for the remainder of that show. This year however was a very different affair, with the D.IV making a spirited appearance during the WW1 scenario.

Top: Sometimes it's just easier to get out and push! **Left**: Settling back to earth amidst the vines at Omaka Aerodrome. **Below**: This photo highlights the 'square box' nature of the fuselage. **Below Left**: Gracefully banking away from the camera. **Bottom Right**: The D.IV features a highly characteristic tail section. **Facing Page**: Dave Horrel piloted the Halberstadt at Classic Fighters 2005. Photo Alex Mitchell.

Airco DH-2

On static display for Classic Fighters 2003, the Geoffrey de Havilland designed DH-2 'pusher' made its airshow debut at this year's show. The aircraft was initially built in the USA, but was brought to New Zealand in 2002 and subsequently underwent a restorative maintenance process, which was completed in late 2003.

Top Right: Sweeping in low over the vineyards, the DH-2 shows that even for a 1916 design, it's a maneuverable aircraft. **Left**: This Kinner radial is not the original engine type, but it still presents some challenges for Roger Harris, who must climb inside the tail booms to spin the prop at startup. **Below Left**: The DH-2 and DH-5 make their airshow debuts together. **Below Right**: Pilot Simon Paul sits out in front of the 'bathtub' fuselage with an excellent forward view. **Bottom**: Touchdown at Omaka. **Facing Page**: The DH-2 above a very French-looking Marlborough. Photo Alex Mitchell.

Airco DH-5

Designed late in 1916 when the DH-2 and other 'pusher' aircraft were still in widespread use, the DH-5 was Geoffrey de Havilland's attempt to design a combat aircraft with the performance of a 'tractor' aircraft, but with a better forward field of vision like the 'pushers'. Succeeding in this goal, the aircraft was not great in combat, and only 550 examples were built before they were retired to a training role in late 1917.

Top: Stuart Tantrum taxis the aircraft to the western end of the field for take off. **Left**: The DH-2 and DH-5 represent the two earliest de Havilland designs still flying in NZ. **Centre Left**: From the rear, the backward stagger of the top wing is not so obvious. **Centre Right**: The business end. **Bottom Left**: As with most early aircraft, the DH-5 features a substantial amount of bracing wire for added strength. **Bottom Right**: During both World Wars, many communities and groups joined together to raise funds to purchase arms, aircraft and ships. **Facing Page**: The DH-5 in its natural element above Omaka. Photo Craig Justo.

20 Classic Fighters Marlborough 2005

Pfalz D.III

This D.III has been resident at Omaka since 2000, but until this year had not flown. Originally built in 1966 for the motion picture *The Blue Max*, the star of that movie, George Peppard, actually flew this machine for over twenty hours in total during the filming. Arriving at Omaka in a somewhat worse for wear state, an exhaustive restoration has been carried out over the past four years, and visitors to this years show were thrilled to see the aircraft finally take to the air.

While all the other First World War aircraft at Omaka wear historically accurate colour schemes representing actual aircraft, the Pfalz's scheme is a little different. Thanks to local man Chris Boyce, the aircraft once again wears the spurious 'seven-colour' lozenge scheme applied by 20th Century Fox during the production of the movie. WW1 German colour schemes only used five colours.

Alex Mitchell

Gavin Conroy

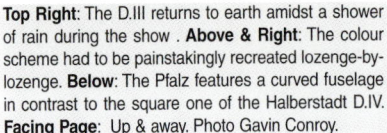

Top Right: The D.III returns to earth amidst a shower of rain during the show . **Above & Right**: The colour scheme had to be painstakingly recreated lozenge-by-lozenge. **Below**: The Pfalz features a curved fuselage in contrast to the square one of the Halberstadt D.IV. **Facing Page**: Up & away. Photo Gavin Conroy.

Philp Merry

Geoff Sloan

Nieuport 24

This aircraft made its New Zealand debut at Classic Fighters 2003, albeit in a different livery. Originally sporting the off-white colour scheme of French WW1 ace Charles Nungesser, the aircraft was repainted in February 2005. The new scheme was copied from a Nieuport 27 (N5246) flown by Sous Lieutenant Gilbert Discours of N87 (a French unit).

A tailskid problem just prior to Easter weekend meant the aircraft didn't fly again during the airshow, but it was nevertheless a treat to see this 'new' Nieuport on the flight line.

Facing page photo: Gavin Conroy.

Geoff Sloan

Geoff Sloan

Lyle Pethig

Reenactment

Those who attended Classic Fighters 2005 were witness to the largest ground theatre displays ever staged in New Zealand. The growth of these reenactments has been almost exponential with every show, and some people have asked why there is such a large element of reenactment at an airshow.

The answer lies in our inaugural show in 2001. During the planning for what was initially thought of as a small but unique airshow, the subject of ground theatre was raised—initially WW1, but expanded to cover both wars. It was during the lead up to the show that ideas for a number of props were discussed, and before long what seemed to be crazy (some would say impossible) ideas were made manifest. While the tanks built for 2001 were merely replicas, people who attended that show were treated to the rare sight of both the British WW1 Male tank (sadly destroyed

soon after the show) and the WW2 German Panther. The enthusiasm of a number of volunteers also meant that a number of troops from both wars were present on the field, along with authentic weaponry and large pyrotechnics. The favour-

able crowd reaction ensured that ground theatre has become a very large part of the show at Classic Fighters, and this is set to continue.

For the 2005 show our brief was to provide an exciting entertainment spectacle, along with an educational component. It was deemed appropriate to highlight the New Zealand contribution to wartime actions, especially within the framework of the French theme.

Once a number of possible scenarios are decided on, the planning goes ahead. This year Warren Mahy coordinated the WW1 scenarios, portraying the New Zealand assault on Les Quesnoy in France, and the death of Baron von Richthofen, while I looked after the WW2 scenarios and the T.E.Lawrence vs Bristol Fighter race.

Unfortunately this year we bit off a little more than we could chew in some areas, and some planned scenarios either didn't take place or didn't happen the way we intended. Logistically a large ground show like this can be difficult to coordinate, and we don't have the luxury of second takes, so unforeseen problems can and do occur. Problems with theatre are to be expected, but we at Classic Fighters accept them as a small price to pay, and will continue to push the boundary of reenactments at our shows. However the other side of the coin is that again we presented something worthwhile to the public. Many of the uniforms being worn, and shown in some of the photographs in this book, will previously only have been seen in wartime photographs and film. At

Omaka the public have the opportunity to study the uniforms, weapons and vehicles at very close quarters, and witness them in action. We should never lose sight of the fact that this is a unique opportunity.

Regardless of the enjoyment of the crowd and reenactors, it is sobering to recall that within a few days of the Classic Fighters show, ANZAC day is celebrated in Australia and New Zealand. This is a time to reflect on the sacrifices our forefathers made in many parts of the world, and as such the final toast at the reenactors evening get together at Omaka is always to the ANZACS—Lest we forget. ❖

Dave Lochead - Ground Theatre Director

Top Left: Jim Lochead as a French 3-star General. **Top Centre**: Some of the WW1 German troops testing their weapons. **Top Right**: Spike Wademan (creator of the poster artwork for this years show), as a British Paratroop officer. **Centre Page**: Allan Mitchell as a WW1 Prussian officer. **Centre Left**: Greg Olsen, who produced many of the fine uniforms seen on the reeenactors at all three airshows, along with Nigel Johnson, as WW2 German troops. **Above**: Part of the WW2 German contingent get ready for action. **Left**: WW2 Allied troops. **Facing Page**: Three reenactors hamming it up for the camera. From left to right, Billy Brittenden, Dave Lochead, and Julian Denmead. Photo Karen Mitchell.

This Page: Members of a number of historical reenactment groups from around New Zealand were at Classic Fighters this year, with their vehicles and weapons. This pastime is growing in popularity, and it's a chance for those involved to portray historical events, while also entertaining and educating the crowd.

Facing Page: A few of the WW1 reenactors try and shoot down a Triplane or two! Photo Phil Teague.

28   Classic Fighters Marlborough 2005

WW1 Reenactment: Le Quesnoy

On 4 November 1918, several days before the Armistice, the small town of Le Quesnoy in northern France was liberated by New Zealand troops from four years of German occupation. The town held 1,500 Germans who refused to surrender. Four hundred troops from the New Zealand division were wounded, 93 of whom died and are buried in Le Quesnoy's cemetery.

The following is part of an account of the battle as told by Lt Harry Henrick, of the New Zealand Rifle Brigade:

"The town was encircled by a huge brick wall, 12-15 metres high, with a 200 metre-wide ditch around the outside, inside of which had been built, at regular intervals, a dozen metre-high brick bastions covered with trees and scrub. These bastions, despite their construction well before modern weaponry, gave the Germans, who were equipped with machine guns, a strategic view point and cover."

"Our advance was covered by a mobile artillery barrage, which was halted by the wall and also by the fact that there were civilians in the town. Nevertheless, at around 5.30am our artillery bombarded the walls and the external defences of the town, by using flaming petrol bombs launched with special guns, driven into the ground. The charge, electrically triggered, exploded in a ball of fire at the moment of impact, causing great terror in the defenders' ranks.

It was impossible to know if this flaming petrol bombardment caused serious losses to the enemy, but it certainly affected their morale. It was a Dantesque spectacle to see the German machine-gunners leaping about frenetically above the wall, their silhouettes standing out perfectly against a wall of flames. And since then, I have never heard of such a beautiful bombardment method. Neither have I heard of this kind of bombardment being used, since then.

"After capturing this section of the front line, our company held fire momentarily in order to help the victims and to transfer the prisoners. After having registered several losses, the companies at the head of the 4th Battalion arrived at the wall.

After hesitating, we could see that by placing the surviving ladder against a narrow wall that covered the inside ditch, it just managed to reach the height of the main wall. Once it was raised for the first time, the Germans prevented us from using it by throwing many stick-bombs joined together in a bundle.

"During this time, Capt Wapier, an officer from Auckland and a mortar fire specialist, bombed the top of the wall with special mortars whilst the others, infantrymen, maintained fire with "Lewis" rifles and standard rifles. Under the cover of this fire, 2nd Lt Averill, assisted by 2nd Lt Kerr and his section, scaled the ladder, losing no time in reaching the top and even entered the town right under the noses of the dumfounded Germans, surprised by such great speed. The Germans took refuge in burrows that they had dug into the ground. The rest of our troops, led by the CO scaled the wall as fast as their rifles, ammunition and other effects would allow them.

"Our first concern on entering the town was to encircle and disarm the prisoners, to extinguish the burning homes - in fact the stables were on fire with horses inside - and to clear the mines and traps. One of the Germans proved to be their paymaster...he had on him the equivalent of two weeks' pay for the entire garrison. At that moment, he was as unaware as us, that in two weeks we would be in Germany where we would receive our pay in marks, the bank notes in marks were to

become mere souvenirs, some of us even used them to light our pipes.

The civilians in Le Quesnoy had been under German domination for four years and their joy literally exploded. Their elation at being liberated was wonderful to see. We were covered in mud, exhausted and marked by the ordeals of the battle. But everyone - women, the elderly, children and even the Little Sisters of the Poor - came out of their cellars to come to meet us, laughing and crying, they gave us kisses, flowers and even food.

Top Right: Kiwi troops on the outskirts of Le Quesnoy before the attack on 4 Nov 1918. Alexander Turnbull Library, Wellington, NZ. Reference: MNZ-1996-1/2; F. Used with permission. **Above**: French troops enjoy some light refreshments before battle commences. **Left**: The Kiwis prepare to advance, with support from the British Mk IV tank. **Facing Page**: A German machine gun post attempts to repulse the Kiwi advance. Photo Philip Merry.

"It was the 4th Battalion that was the first to arrive in the town... but they were upstaged by the 2nd Battalion who marked their arrival with their very own orchestra. When the French flags appeared and they played the French national anthem in the square, most of the elderly were crying uncontrollably.

Gavin Hadfield

Erin Boyd

"This is how our final battle ended. It received a lot more attention than previous battles because it was unique in many ways: the attack of an old walled city occupied by civilians; flaming petrol bombs; the use of ladders to scale the walls; all of this was an unusual mix: a war both medieval and modern. The result was so much more than satisfactory.

"All of this meant so much more than the conquest in Passchendaele of 600 metres of muddy ground, pitted by shells, especially when you consider the fact that not one civilian life was lost in the entire operation!" ❖

Erin Boyd

Philip Merry

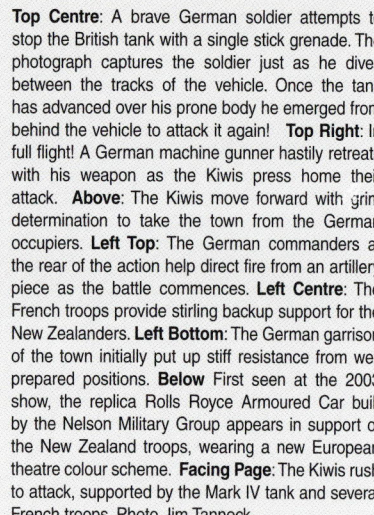

Top Centre: A brave German soldier attempts to stop the British tank with a single stick grenade. The photograph captures the soldier just as he dives between the tracks of the vehicle. Once the tank has advanced over his prone body he emerged from behind the vehicle to attack it again! **Top Right**: In full flight! A German machine gunner hastily retreats with his weapon as the Kiwis press home their attack. **Above**: The Kiwis move forward with grim determination to take the town from the German occupiers. **Left Top**: The German commanders at the rear of the action help direct fire from an artillery piece as the battle commences. **Left Centre**: The French troops provide stirling backup support for the New Zealanders. **Left Bottom**: The German garrison of the town initially put up stiff resistance from well prepared positions. **Below** First seen at the 2003 show, the replica Rolls Royce Armoured Car built by the Nelson Military Group appears in support of the New Zealand troops, wearing a new European theatre colour scheme. **Facing Page**: The Kiwis rush to attack, supported by the Mark IV tank and several French troops. Photo Jim Tannock.

Chris Guy

Erin Boyd

Karen Mitchell

32 Classic Fighters Marlborough 2005

Pither Monoplane

In 1910, pioneer New Zealand aviator Bert Pither designed and built a Bleriot-style monoplane which he claimed to have flown at Oreti Beach (Southland) on the 5th of July that year. Unfortunately there were no other witnesses to these claimed flights, and so the question of whether the aircraft actually flew or not has remained unverified ever since.

Colin Smith of the Croydon Aircraft Company (Mandeville) has built a replica of the aircraft to test the design. Working drawings were produced from contemporary newspaper reports and photos, while Bill Sutherland (Waikaka) built a look-alike Pither V-4 engine to power the craft. As identical to the original as possible, the replica is 77kg heavier than Pither's original design, but as witnessed by attendees at this years show, is still capable of making it into the air under its own power.

Paul Charal

Top: Pilot Jerry Chisum waves to the crowd as he taxis the aircraft back to its hangar after a successful display. **Centre**: The minimalist design of the Pither is very much in keeping with other aircraft designs of its day. **Bottom**: Appropriate branding on the tail! **Facing Page**: The Pither makes a short hop at Omaka. Photo Alex Mitchell.

Gavin Hadfield

Mike Hodgkinson

ASHTON TECHNOLOGIES

Contact: Des Ashton
Ph: + 64 3 577 6972
Mobile: + 64 21 664 920
Email: des.ashton@ihug.co.nz
Address: 73 Morven Lane, Fairhall, RD2, Blenheim 7321, N.Z

ASHTON TECHNOLOGIES LIMITED

AVIATION & DEFENCE INDUSTRY CONSULTING & SERVICES

* Major Tender Development
* Product Representation
* Strategic, Marketing, Management and Training Advice
* Project Management
* Temporary management assignments

34 Classic Fighters Marlborough 2005

Brough Superior Motocycle & The Race

At each Classic Fighters show we try to present a feature vehicle—in 2001 it was the Panther tank, while 2003 marked the arrival of the Weta Workshops-constructed WW1 Male tank and the display of the Southward Car Museum's Mercedes 770k staff car. For 2005 we planned to portray something very special—an event that had not been recreated accurately since it took place in the early 1930s, and at the same time draw public attention to the legendary figure of T.E. Lawrence (of Arabia).

During Lawrence's stint in the RAF in the '20s and '30s, he owned no fewer than seven Brough Superior motorcycles. In his second book, *The Mint*, published under his then assumed name of T.E. Shaw, he wrote a passage entitled 'The Road', describing an encounter and subsequent 'race' he had with a Bristol Fighter F2b aircraft, while riding one of his Broughs (*pronounced 'bruff'*).

A great deal of planning was involved in this event, not the least of which was finding a Brough Superior, widely considered to be the peak of 1930s motorcycle development, and an owner who would be happy to let us use his machine. Luckily Ian Nielsen from Auckland came to our rescue.

For many people this would be the first, and probably only time, that they would witness such a rare motorcycle being ridden, and for myself, it was a rare privilege to ride it.

Traditionally Marlborough's Easter weekend weather is good, but this year was an exception. Good Friday is generally the practice day for the show when we iron out problems such as timing ground events with the air operations and working our way through any issues that arise. Those who were present on Friday will remember that the weather was shocking—very cold with heavy rain. In these conditions we couldn't practice, so Saturday became the practice day. By noon the air operations were running 10 minutes ahead of schedule, and when the Brough took the field at the appropriate time the Bristol had already landed. On Sunday, when every-

thing was on time and proceeding as planned, the Bristol Fighter's engine developed a major problem and was not able to take to the air, hence the race reenactment could not be run. The disappointment felt by both the organisers and crowd was understandable, but unfortunately these problems do arise from time to time with old & rare machines—usually when you least want them to. ❖

Dave Lochead, Ground Theatre Director

Right: The peak of 1930's motorcycle development. **Below:** Practicing for, and filming the Brough vs F2b race prior to the show. **Bottom:** Dave Lochead recreates a scene from 'The Road'. **Facing Page:** Leading Aircraftsman Shaw, and his trusty steed at Timara Lodge (Blenheim) during filming for 'The Road' TV news item. Photo Erin Boyd.

Percival Proctor V

In keeping with the goal of providing even more entertaining and educational theatrics this year, we were pleased to be able to recreate the arrival of New Zealand aviatrix, Jean Batten, at Omaka aerodrome in 'her' Percival aircraft (in this case ZK-AQZ owned by the Masterton-based Sport and Vintage Aviation Society). In October 1936 Jean broke the England-Australia solo flight record when she reached Darwin in 5 days 21 hours in her Percival Gull Six, which was a forerunner of the Proctor series of aircraft. After flying on to New Zealand she toured the country in her aircraft, and included a stop at Omaka Aerodrome. Needless to say, we were very pleased to have 'Jean' and her aircraft back for this year's show!

Top: 'Jean Batten' (a.k.a. Kirsty Barrett-Hamilton) waves to the crowd after arriving at Omaka. **Below & Right**: The beautifully restored SVAS Proctor V first flew again in 1993. **Bottom**: This aircraft was first flown to New Zealand by Percival agent, Ernle Clarke, in 1949. **Facing Page**: Photo Paul Maggs.

de Havilland Airliners

de Havilland has always been a name synonymous with aviation in New Zealand, so this year we were very pleased to see three of the workhorses of early New Zealand passenger services at Omaka for the show. All three aircraft, the DH-83 'Fox Moth', DH-89B 'Dominie' and DH-90A 'Dragonfly' are based with the Croydon Aircraft Company at Mandeville in Southland.

Bert Mercer's Air Travel (NZ) Ltd operated ZK-ADI and at least one other Fox Moth on the West Coast of New Zealand for a period during the 1930s. Between 1943-1948 this aircraft was also pressed into service with the RNZAF. Air Travel also operated several Dragonflys, one of which tragically crashed into the sea. Other civil operators such as the National Airways Corporation (NAC) and Air Charter (New Zealand) Ltd were among many pioneering companies who used these de Havilland aircraft to open up the New Zealand air routes.

Top: The DH-89B 'Dominie' wears the colours of the National Airways Corporation, with which it flew in the 1930's and 40's. **Centre**: Four Gipsys in harmony as the two twin-engined de Havillands fly by. **Above**: This DH90a 'Dragonfly' is one of two intact surviving examples from the 66 originally built. **Below**: This head on view clearly shows these aircraft are closely related. **Facing Page**: The Fox Moth races past. Photo Philip Merry.

Beech Model 17 'Staggerwing'

The Beech Staggerwing never fails to turn heads wherever it goes around New Zealand, and Omaka this year was no different. This stunning epitome of 1930s aviation elegance is owned and flown by Robin Campbell of Auckland. This aircraft is one of two Staggerwings now resident in New Zealand.

Geoff Sloan

Paul Charal

Mike Hodgkinson

This Page: Four views of this beautiful aircraft which illustrate just why it is a crowd favourite. **Facing Page**: The sleek Staggerwing taxis past the crowd line after it's display. Photo Paul Charal.

Geoff Sloan

Lockheed Electra 12a Junior

Arriving in New Zealand just a few weeks too late to appear at Classic Fighters 2001, this elegant aircraft always puts on a great display. Permanently based at Omaka Aerodrome and owned by Pat Donovan, the 12a Junior is a smaller (six-seat) version of the larger 10-seat Lockheed Electra 10s used on the civil passenger routes in New Zealand between 1937 and 1950. Using similar Pratt and Whitney nine-cylinder radial engines as the North American Harvard, some pilots have said that sitting in the Electra's cockpit is like having a Harvard strapped to each knee—and Pat certainly knows how to make the engines growl when flying a display routine!

Paul Maggs

Mike Hodgkinson

This Page: No matter which way you look at the Electra, it's a beautiful looking aircraft, and it's always great to see her flying about the Marlborough skies, even if they were a little gloomy at Easter. **Facing Page**: Twin engined and twin tailed elegance as the L12a lands. Photo Lawrence Ackett.

Alex Mitchell

Mike Hodgkinson

BASIS
For All Your Aviation & Auto Accessories

- Silk scarves
- Flying goggles
- Leather flying helmets

Plus much more, available from:

31 High Street, Renwick 7352, Marlborough, New Zealand
Phone: +64-3-572-8880   Fax: +64-3-572-8851
Web: www.basisnz.com   E-mail: sales@basisnz.co.nz

44 Classic Fighters Marlborough 2005

Fleet 16b Finch

Unfortunately pilot and owner Graham Orphan had many duties to perform at Classic Fighters 2005, including working on the PA system, thus his delightful Fleet 16 couldn't take part in the main display part of the airshow. However, as illustrated on these pages, the aircraft did get out and about, and was displayed in the aircraft park for everyone to enjoy. The only other Fleet aircraft to have been operated in Australasia was a Fleet 7b operated by the Marlborough Aero Club in the late 1930s, so the arrival of this aircraft at Omaka in 2004 was very appropriate.

This Page: The black, yellow and red colour scheme creates a striking aircraft, while the Kinner radial engine is a delight to listen to. **Facing Page**: Owner/pilot Graham Orphan (up front), and photographer Craig Justo take the Fleet out for a spin just prior to the show. Photo Alex Mitchell.

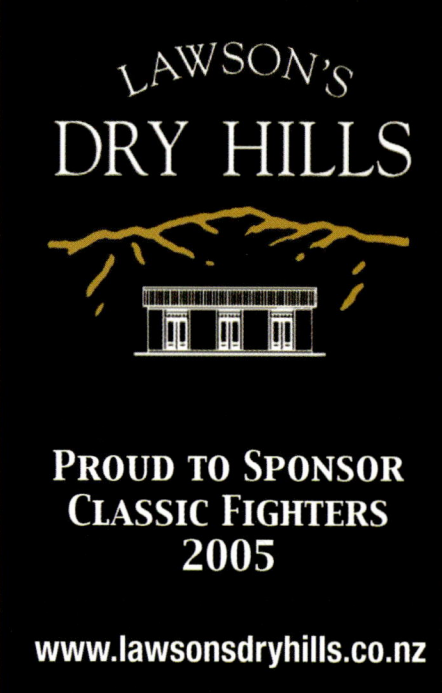

Other Sights

There's always more than just aircraft to see at a show the size of Classic Fighters—many special and memorable sights were captured by our photographers over Easter.

Above: A gathering of old, and not so old Corsair pilots next to the ex-RNZAF FG-1D Corsair on Saturday. Left to right; Gene de Marco, Keith Skilling, Geoff Thorpe, Gordon Neill, Peter Keddell, Leyton Stephenson, James Slade (current owner of the aircraft), Skip Watson, Laurie Cromie, Trevor Foley, Bill Newfield, Henry Eccersall, Neil Manssen. Front: Sir Tim Wallis.

Above: Former world champion glider pilot Ray Lynskey always puts on a superb, albeit silent, aerobatic display in his Ventus glider. Here he swoops along the crowd line prior to landing.

Below Left: Flames belch from the P-51 exhaust stack as the engine is fired up.
Below: The show not only includes classic aircraft. **Below Centre**: If we're having old aircraft at the show, we may as well have some 'old' firemen as well.

Bottom Left: The NZ Fire Service are always on hand 'just in case'. In previous years we have had grass fires on the aerodrome. **Bottom Right**: And if we're going to bother to have firemen, then we may as well have some big fiery blasts!
Facing Page: David Marwick (front) and Graham Orphan reposition the Omaka syndicate Tiger Moth ZK-BER to the aircraft display park. Photo Paul Maggs.

Top Left: Mark Helliwell and Mishelle Lawson took first prize in the Classic Fighters Costume Competition, and won themselves a weekend at Parautane Lodge near Nelson. **Above**: Stephen Witte's L-19 BirdDog flew extensively during the weekend as a camera ship for some of the air-to-air photo shoots. **Left**: 'A Few Good Men'—dressed as US Navy fliers, in authentic USN uniforms. **Below**: With a French theme, it's not too surprising we found at least one mime artist in the crowd. **Below Right**: Leigh Dixon in her hand-made replica 1860s French ball gown. **Bottom:** Classic Fighters 2005 was the first airshow in New Zealand to make use of big screen technology for the benefit of the crowd. **Facing Page:** Lester Hope's awe inspiring Eiffel Tower replica. Photo Karen Mitchell.

50 Classic Fighters Marlborough 2005

Static Aircraft

While Classic Fighters is about flying aircraft, the fact remains that there will always be aircraft which for one reason or another have reached the end of their flying life. We prefer to think of these as future restoration projects, rather than old has-beens, so it's not surprising that there are always a few earthbound aircraft to see at the show as well.

Right: The RNZAF brought BAC Strikemaster NZ6375 across from Base Woodbourne—it last flew in 1991. **Below Left**: Finished in 1940, this Lincoln Sport is the oldest home-built aircraft still in existence in NZ.

Above Right: The RNZAF's Bell 47 'Sioux' helicopter from the Ground Training Wing at Base Woodbourne. **Left & Below**: The Bristol Freighter 'Merchant Courier', an Omaka icon, is one of the aircraft flown on regular services across Cook Strait by Safe Air between 1951-1986. **Facing Page**: The jewel of the Marlborough Warbirds Assn is this (non-flying) Hawker Hurricane replica originally built for the 1969 movie *The Battle Of Britain*, and restored by Ron Flintoft and his team. Photo Karen Mitchell.

Boeing 757 - RNZAF

The RNZAF's No 40 Squadron currently operates two Boeing 757 aircraft as long-range military and VIP transports. Entering service in late 2003, this was the first occasion that this type has displayed at Classic Fighters, and after the noisy display that was put on for the visiting crowds, hopefully won't be the last—there's nothing like a low level flyby of a jet aircraft of this size!

Above: The latest RNZAF aircraft to enter service seen approaching one of the original New Zealand Permanent Air Force types, the Bristol Fighter F2b. The NZPAF was the forerunner of the RNZAF, and operated a number of F2bs during the 1920's. **This Page**: Low, slow and very noisy, the 757 put on an impressive display for the crowd. **Facing Page**: The Lockheed Orion peels off while the Boeing 757 begins its solo display routine. Photo Jim Tannock.

CT-4 Airtrainer - RNZAF

The RNZAF's 'Red Checkers' display team is made up of volunteer members (instructors) from the Air Force's Central Flying School, and always puts on a very polished display. When the team was first formed in the 1960s the aircraft used was the North American Harvard, which was the current air force trainer. By the time the team was re-formed in 1980, the New Zealand designed and built Airtrainers had become the standard primary trainer, and this aircraft has been used by the team ever since (albeit in upgraded models).

Above: Is that one or two? **Left**: Pulling up into the top of a five ship loop. **Centre Left**: A massed formation take off prior to the display. **Centre Right**: The 'mirror formation' is one of the team's signature routines. **Bottom**: The final act of the display is the inward break in which all aircraft other than the leader banks inwards toward the rest of the team—a tricky maneuver which looks stunning from the ground. **Facing Page**: Coming down the other side of a loop. Photo Paul Maggs.

Gavin Hadfield

Lockheed P3 Orion - RNZAF

Operated by the RNZAF since 1966, upgrades for the fleet of six Orions are expected to increase its lifespan until at least 2015. As well as purely military roles, the P3 also regularly patrols New Zealand's 200 mile Economic Zone, and often carries out vital search and rescue work in the South Pacific region. Many a distressed yachtsman has been very relieved to see the shape of an RNZAF Orion coming into view over the horizon! With the RNZAF's C-130 Hercules fleet otherwise occupied, this year the Orion was the only four-engined aircraft to display at Classic Fighters. That aside, the crew put on a terrific handling display which everyone thoroughly enjoyed. Now, if we could just get two of them to display together for the next show...

Above: Coming in low and fast from the west, this part of the display also excites the crowd. **This Page**: Other views of the 'slow' section of the handling display. **Facing Page**: A great view of the top of the aircraft as it banks around the curved crowd line at Omaka. Photo Philip Merry.

58 Classic Fighters Marlborough 2005

Bell UH-1H Iroquois / Kiwi Blue - RNZAF

As usual the RNZAF also provided entertaining displays from the Bell UH-1H Iroquois helicopter, and the 'Kiwi Blue' Parachute Display team.

The Air Force's No 3 Squadron has operated 16 Iroquois since 1966, and they continue to play an integral part in the service's operations. Used primarily in Army support roles, the Iroquois also plays a huge role in Search and Rescue operations throughout New Zealand.

The Kiwi Blue team is comprised of instructors from the Air Force's Parachute Training and Support Unit. Since 1964 this unit has trained personnel from all sections of the New Zealand Defence Forces, and in some cases foreign troops as well.

Philip Merry

Gavin Conroy

Above Left: A German soldier shoots at the descending Kiwi Blue 'paratroopers' at the start of the WW2 ground theatre action. **Above Right**: The Iroquois runs in low and fast from the south prior to starting its display routine. **Left and Below**: At the start of the WW2 ground theatre scenario, the Kiwi Blue 'paratroopers' make the mistake of dropping in right into the midst of the German troops, who immediately move to ensure the paratroopers are quickly 'neutralised'. **Bottom**: The Huey lifts off from the western aircraft park before displaying. **Facing Page**: The Iroquois crew displays the technique for transferring someone to and from a moving vehicle. Photo Gavin Conroy.

Gavin Conroy

Gavin Hadfield

Paul Maggs

60 Classic Fighters Marlborough 2005

Restoration Row

There have always been pieces of old aircraft hidden away in the many private hangars at Omaka Aerodrome which are one day destined to fly again. Sometimes that day is not far away, and at other times, it may be a very long time into the future. As at previous shows, this year we dragged a few of the aircraft projects out of their hangars to show everyone the progress that has been made, and to tantalise you with the aircraft that you might one day see in the air at a future Classic Fighters show.

MARLBOROUGH AERO CLUB

Address: Omaka Airfield,
Aerodrome Rd,
Blenheim
Phone: (64) 3 578 5073
Fax: (64) 3 578 1817
Email: marlbaero@xtra.co.nz
Web: www.marlboroughaeroclub.co.nz

We like to share our passion for aviation so choose us for:

* flight training and type ratings
* aircraft hire
* scenic flights
* picnics at the beach
* vintage aircraft experiences
* aerobatic rides

Top Left: Graeme Frew's Yak-3 is actually in far better shape than it appears here, and may fly again within a couple of years. **Left**: The local Spezio Tuholer. **Below**: Mike Nicholl's Finnish Air Force Curtiss Hawk 75 was shot down by Russian aircraft in 1941. **Facing Page**: An original Royal Aircraft Factory BE2 being restored to airworthy condition. Photo Karen Mitchell.

Last Flight Of The Red Baron

On April 21 1918, Manfred von Richthofen, often referred to as the 'Ace of Aces', was shot down and killed over Morlancourt Ridge, near the Somme River. Von Richthofen had been pursuing a Sopwith Camel flown by a Canadian pilot, Lt Wilfrid May, when he in turn was chased by fellow Canadian Capt Roy Brown in a second Camel. Throughout the chase and subsequent dogfight, von Richthofen's triplane was also shot at by a number of Australian troops on the ground, with both rifle and Lewis machine gun fire, as the aircraft flew overhead.

Von Richthofen's red triplane came to rest in a field near the village of Vaux-sur-Somme, in an area controlled by the Australians. Some accounts of the final moments of his life claim he was already dead when his plane came to rest, while others say he was still alive when the aircraft crash landed, but died almost immediately when the aircraft came to a complete stop.

Irrespective of when he died, a subsequent examination of his body showed that he was killed by a single bullet which entered the right hand side of his body near the armpit, and exited at a higher level from the left side of his chest.

Geoff Sloan

While Brown was initially credited with shooting down von Richthofen, and was awarded a bar to his DFC for his efforts, it is now generally accepted that the fatal shot was probably fired by one of the Australian troops. The most likely candidate was Sgt Cedric Popkin of the Australian 24th Machine Gun Co, although in the end it is unlikely to be proven beyond doubt.

The British arranged a full military funeral for von Richthofen, and he was buried near Amiens on April 22nd. Six RAF officers acted as pallbearers while an Australian honour guard fired a salute. After the war, his remains were exhumed and reburied in the Richthofen family cemetery in Wiesbaden, Germany.

Gavin Conroy

Philip Merry

Alex Mitchell

Top: 'Capt Brown' chases von Richthofen, **Above:** The Baron's Dr.1 'crash lands' and Australian troops race to the scene. **Far Left**: On removing the body from the aircraft the Australians realise the pilot is dead. **Bottom Left**: The Baron's body is removed from the crash site. **Left**: John Lanham, aka The Baron, wearing the actual scarf that was removed from Richthofen's body on 21 April 1918. **Below**: Members of No 3 Sqd Australian Flying Corps around the fuselage and tail section of what remains of von Richthofen's aircraft after souvenir hunters had finished with it. **Facing Page**: The Baron's Dr.1 in all it's glory. Photo Jim Tannock.

Erin Boyd

Cri Cri

After we had decided to have a airshow with a French theme, we then had to work out if we could have any French aircraft appear. Given that French aircraft are somewhat few and far between in New Zealand, we jumped at the opportunity when Neville Hay offered to bring and display his French-designed Cri Cri homebuilt—the smallest twin engined aircraft type in the world.

The name 'Cricket' is used in some English speaking countries to refer to the type (hence the artwork on the tail) but it's not a good translation of Cri Cri which literally means the noise (chirping) made by a Cricket.

Paul Maggs

Geoff Sloan

Gavin Conroy

Far Left: The aircraft is powered by two two-stroke engines providing 15hp each. **This page**: These photos illustrate the diminutive size of the CriCri—the aircraft is only 3.9m long, 1.21m high and has a wing span of 4.88 metres. When empty the aircraft weighs only 87kg. **Facing Page**: As the Cri Cri takes to the air it's easy to see that pilot/owner/builder Neville Hay would not want to be a taller man. Photo Geoff Sloan.

Geoff Sloan

Percival 'Piston' Provost

This Provost, owned and operated by the Masterton-based Old Stick and Rudder Company, is one of two such aircraft currently resident in New Zealand. While not previously used in this country, the Provost is a direct ancestor of the BAC Strikemaster, which was operated by the RNZAF for many years. This OSRC aircraft was painted in a temporary scheme for Classic Fighters 2003, but it now permanently wears the same Sultan Of Omans Air Force colours. While the Provost was used primarily as a trainer in British service, in Oman the aircraft was fitted with machine guns and rockets, and was used as a light ground attack aircraft with some success.

Alex Mitchell

Gavin Conroy

Gavin Conroy

Mike Hodgkinson

Geoff Sloan

Left: The Provost, one of the RAF's training aircraft, taxis alongside another long-lived Commonwealth trainer, the de Havilland DHC-1 Chipmunk. **This page**: Some of these photographs highlight the 'wide-bodied' nature of the Provost. In contrast with many of the earlier (slim) British tandem trainers, the Provost was designed to seat both pupil and instructor side-by-side. **Facing Page**: Ex-RAF pilot Barry Stott flew Provosts during his time in the Royal Air Force, and now enjoys the chance to put the aircraft through its paces once again. Photo Geoff Sloan.

Thunder Mustang

The Thunder Mustang is a 3/4 scale replica of the North American P-51D Mustang fighter from World War 2. But it's actually more than that—using a specially developed Falconer V-12 engine this aircraft can out-perform the original Mustang. This aircraft ('Red Honey') was imported into New Zealand in early 2005, and is one of only nine Thunder Mustangs worldwide. Being the first aircraft of its type in this country, Civil Aviation regulations mean that a certain amount of 'sedate' flying must be conducted before the aircraft can perform aerobatics. As such the display at Classic Fighters was relatively laid back, but visitors to the show were left in no doubt that when this aircraft has the chance to really 'let loose', it will be spectacular!

Alex Mitchell

Gavin Conroy

Above: The Thunder Mustang sits alongside its larger, and older brethren on the display line. While the smaller size of the aircraft is obvious from this shot, it's also easy to see from the shape of the aircraft that it is an accurate scale replica.

Geoff Sloan

This Page: From whichever angle you choose to view it, the Thunder Mustang is very much as elegant an aircraft as the original P-51—it just looks right! **Facing Page**: The Thunder Mustang takes to the air for its first display in the South Island. Photo Geoff Sloan.

Paul Maggs

Geoff Sloan

Gavin Conroy

Yakovlev Yak-52

Over the past few years the Yak-52 has become a very popular 'warbird' and aerobatic aircraft in New Zealand (it can withstand aerobatic maneuvers of up to 7g)—so much so that there are now around a dozen of these aircraft in the country. First produced in 1976, the '52 was designed as a primary trainer for Eastern Bloc pilots who would later move on to fly Soviet jet aircraft. Over the past 18 months a group of very skilled and dedicated pilots, Brett Emeny, Bryan Coppersmith, Paul Hughan, John Parker and Kevin Jane, have been working hard at becoming this country's first five-ship Yak-52 aerobatic display team. After a stunning display at Classic Fighters 2005, it's obvious that they have well and truly reached that goal, and it seems that the 'Red Star' team name is destined to become a New Zealand icon up there with the 'Roaring Forties' and the 'Red Checkers'.

Top Left: As a four-ship formation, the Yaks make good use of their recently installed smoke kits. **Below Left**: The three-ship mirror formation is rapidly becoming one of the team's signature routines. **Centre**: All six Yak-52's and three Chinese Nanchang CJ-6's at this year's show fly by in a mass formation. **Bottom Left**: From the top, Paul Hughan, Bryan Coppersmith and Brett Emeny, form the three-ship section of the team. **Bottom Right Thumbnails**: Four of the team's aircraft, from the top, Paul Hughan's ZK-ZAY, Brett Emeny's ZK-YAK, John Parker's ZK-TYS, and Kevin Jane's recently arrived ZK-YAC. **Facing Page**: Ex-RNZAF pilots Bryan Coppersmith in the yellow ZK-YNZ, and Paul Hughan get airborne. Photo Chris Guy.

Nanchang CJ-6a

While there have always been a significant number of interesting vintage aircraft based at Omaka Aerodrome, it has to be the Chinese Nanchang that can be credited with being the catalyst for much of the 'warbird renaissance' that has happened here over the past nine years. It was in 1996 that a group of local enthusiasts got together, with the primary goal of purchasing a CJ-6a to be based and operated from Omaka. That one Nanchang quickly became two (ZK-STP 'China Doll', and ZK-WOK '42'), and they generated enough interest in the local community that the Marlborough Warbirds Association was formed. Since then warbird activities have progressed from informal fly-ins through to the highly polished airshow that has become Classic Fighters Marlborough, and we now have literally dozens of warbird and vintage aircraft based on the field, either flying or being restored.

This Page: The two Omaka Nanchangs were joined by the Ardmore-based ZK-OII ('26') this year. **Facing Page:** Barry Stott and Jay McIntyre taxi out past the 'sea of green' in ZK-WOK. Photo Gavin Conroy.

Soko Galeb G2-A

Pat Donovan's sleek Yugoslavian 'Seagull' was the only jet warbird to display at Classic Fighters this year (with the exception of the RNZAF's Boeing 757). After arriving in the country in 2004 the aircraft required a significant amount of restorative maintenance, and this was completed only a few short weeks before this year's show. The test flying from RNZAF base Woodbourne went without a hitch, and the aircraft was able to make its New Zealand airshow debut at Easter. As with the Vampire and Fouga Magister in 2001, the Galeb operated off the hard tarmac at Base Woodbourne, rather than trying to operate from the grass runways at Omaka.

This Page: The Galeb is a sleek aircraft initially used for training, but later modified for a light ground attack role as well. **Facing Page:** Propelled along by the Rolls Royce Viper II turbine engine, Sam Richardson and Peter Turvey scream past the crowd in the fastest aircraft at this year's show. Photo Alex Mitchell.

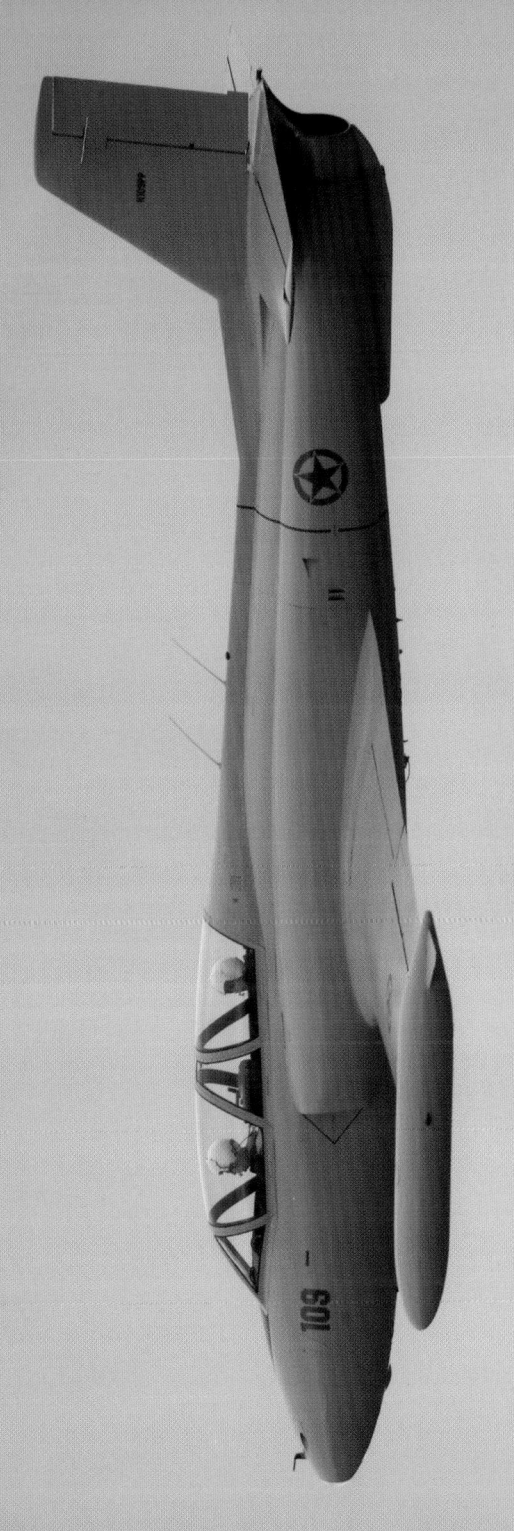

de Havilland DHC-1 Chipmunk

Like the Yak-52, the Chipmunk has become a very popular warbird in New Zealand over the last decade—there are now well over a dozen examples of this type in the country. This year we were privileged to have eight 'Chippies' gathered together at Omaka, and to be able to see them all in the sky at the same time. Most airshow visitors are used to seeing the Chipmunk in 'fly-by' mode, and don't realise that this little aircraft is a very able aerobatic performer as well. This was well illustrated at the show this year by Ralph Saxe flying ZK-JIT—pilot and aircraft put on a sizzling aerobatic display for the crowd prior to the official start of the airshow.

Top Right: Two of the Chippies, including the yellow Canadian example ZK-CVM, take to the air.
Right: All eight Chipmunks took part in the massed formation fly past. **Bottom:** Ralph Saxe in ZK-JIT sweeps past the crowd during his aerobatic display. **Below Left & Right:** The nose profile of these aircraft clearly identifies them as de Havilland designs.
Facing Page: This Old Stick and Rudder Company DHC-1 was used by Prince Philip to learn to fly. Photo Gavin Conroy.

Air Tractor AT-402B

Agricultural aviation in New Zealand was highlighted at this year's show when Masterton-based John Bargh flew a stunning aerial top dressing display routine in his bright yellow turbo-prop powered Air Tractor 402B. Being able to carry a load in excess of 2000kg, and powered by a 680 horsepower engine, this is one heck of an aircraft, and John certainly left the crowd in no doubt that piloting top dressing aircraft can obviously be an exciting career choice.

Gavin Conroy

Lawrence Ackett

This Page: The art of good spraying is to make sure the load goes exactly where you want it to! **Facing Page**: It's difficult to tell if the aircraft is landing, or just making a low fly-by—either way, John's display routine certainly captured the crowd's attention. Photo Gavin Conroy.

Geoff Sloan

Gavin Conroy

Gavin Conroy

Geoff Sloan

Aerobatic Displays

An airshow like Classic Fighters just wouldn't be the same without some high-speed, high-G, single aircraft aerobatic displays. This year we were privileged to see three of New Zealand's best aerobatic pilots take to the skies and show us what they and their machines are really capable of.

Chris Guy

Gavin Hadfield

Top & Bottom Left: Steve Taylor in his Edge 540. Steve has a number of NZ Aerobatic Champion titles to his name, and when flying the 350hp Edge, certainly puts on a great display. **Left & Below:** Richard Hood, who also has many Aerobatic Championship titles under his belt, has been flying the Pitts Special for a number of years, and is often seen at airshows around New Zealand. **Bottom Right**: Doug Booker also puts on a great display in his brightly coloured Giles 202. **Facing Page:** As Doug taxis in after a great display one question remains unanswered—who is the biggest 'NUT'? Is it Doug or the aircraft? Photo Philip Merry.

Philip Merry

Gavin Conroy

Gavin Conroy

82 Classic Fighters Marlborough 2005

Saturday Evening Dinner

This year, to add a little extra to the show, in addition to the day-time airshow, we also offered visitors to Omaka the chance to have a French-style three course evening meal on the aerodrome. This was preceded by a twilight aerobatic display and a hot air balloon 'night glow', and was accompanied by a somewhat risque, Moulin Rouge-themed floor show. If you were there, then you know what we mean...

Philip Merry

Philip Merry

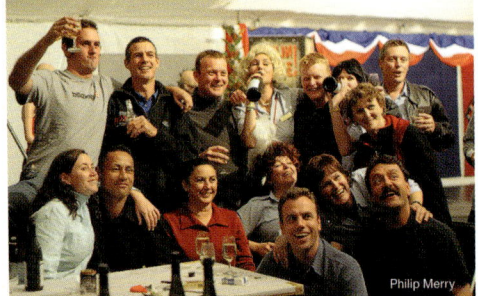

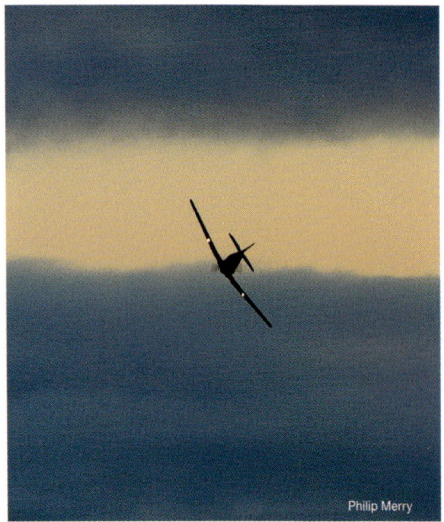

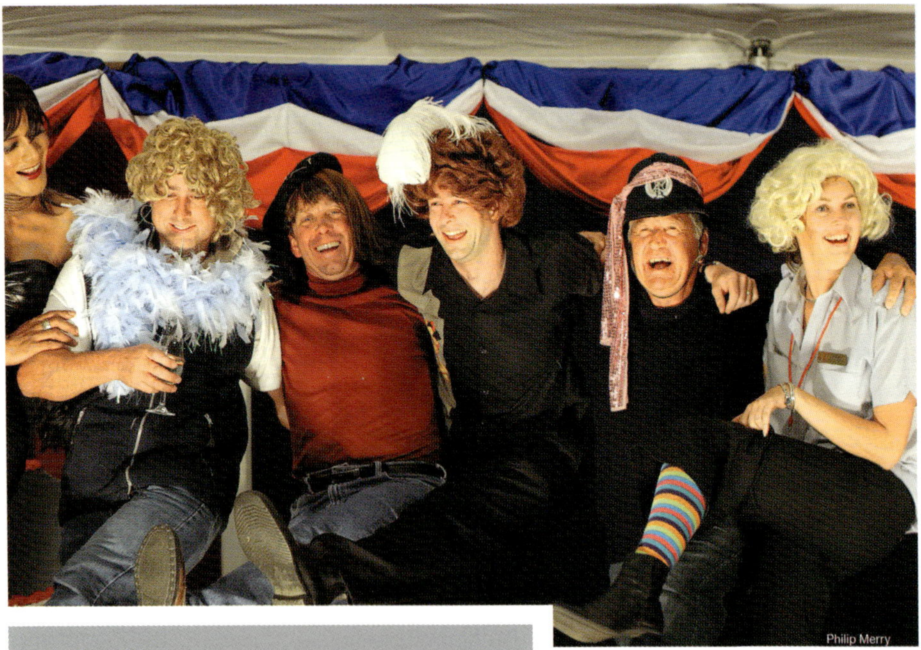
Philip Merry

Top Right: The Harvard put on a great twilight aerobatic display. John Lanham flew the display on Saturday evening while Martin Burdan flew a similar display on Sunday evening. **Top Left & Above:** These photos attest to the fact that our visitors had a good time, and certainly appreciated the fine Marlborough wine on offer. **Facing Page:** The balloon night glow kept our guests intrigued while they enjoyed their pre-dinner drinks. Photo Phil Teague.

Consolidated PBY5a Catalina

The amphibious Catalina flying boat is often seen at shows around New Zealand in her usual guise of a WW2 RNZAF aircraft from No 6 Squadron. This year however, the Catalina Group (syndicate owners of the aircraft) excelled themselves by 're-branding' the aircraft in a striking new, albeit temporary, French Fleet Air Arm (l'Aéronautique navale) colour scheme for the show. The French Navy used many Catalinas in a variety of roles between 1943 and 1971, and it was great to see this big bird acknowledging that side of its' history as well. Luckily there were no ships close at hand that this 'French bomber' could have a go at sinking over the weekend!

This Page: The Catalina is always a crowd pleaser as it gracefully displays its lines and shows how well it handles for a large aircraft, but this year it certainly was 'Magnifique!' **Facing Page**: Floats up! The wingtip floats are retracted as the amphibian cruises past. Photo Philip Merry.

Douglas DC3 Dakota

The second WW2 era big bird to display at this years show was the Warbirds Dakota—ZK-DAK. Outwardly representing the aircraft flown by Kiwi pilot Sqd Ldr Rex Daniell during both the D-Day and Arnhem operations, inside the aircraft is a little more plush than the original. This feature was enjoyed by many visitors to Classic Fighters, who had the chance to take a memorable joyride in this wonderful old warbird.

Mike Hodgkinson

Geoff Sloan

This Page: Flying, taking off, landing or taxiing by, the Dak always evokes memories of days gone by. This is especially true when one remembers that these aircraft were once used extensively throughout NZ. **Facing Page:** For a relatively big aircraft, the Dakota can leave the ground quite quickly. Photo Geoff Sloan.

Mike Hodgkinson

Paul Maggs

D.L.A. Turner

North American Harvard

A regular crowd pleaser at airshows up and down the country, the Harvard has had a long and distinguished career in New Zealand. The throaty roar of the Pratt and Whitney engines is the signal that the Harvard Display Team is about to take to the air for one of their excellent aerobatic displays. It was the RNZAF's disposal of their Harvards in 1977 that prompted the formation of the of the New Zealand Warbirds Association.

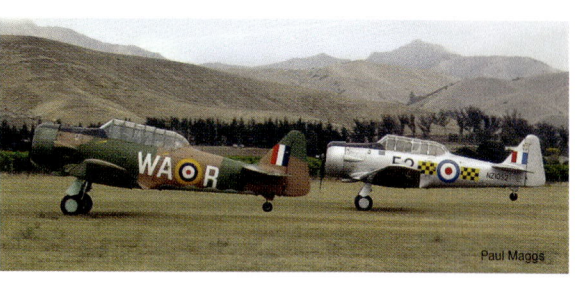

Paul Maggs

Philip Merry

Gavin Conroy

Top: ZK-WAR was one of the first privately owned Harvards in NZ, and it's gratifying to see that she's still flying, and thrilling crowds over 25 years later. **This Page**: There are around a dozen Harvards still flying in New Zealand, and they take every opportunity to get together. **Facing Page:** Harvard '98' roars past the crowd during the NZ Warbirds Harvard Display Team display routine. Photo Jim Tannock.

Geoff Sloan

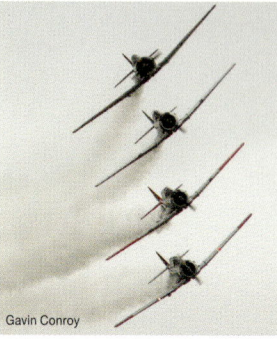

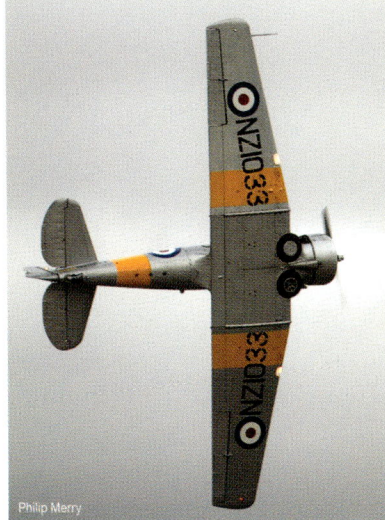

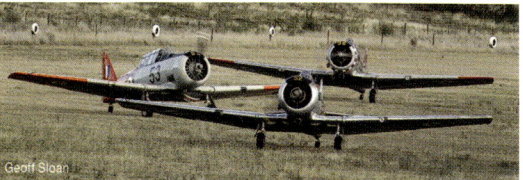

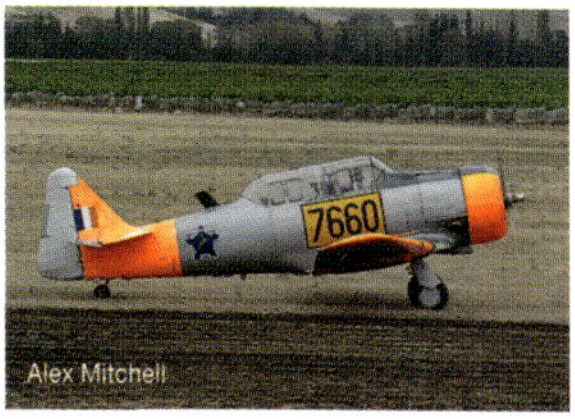

Alex Mitchell

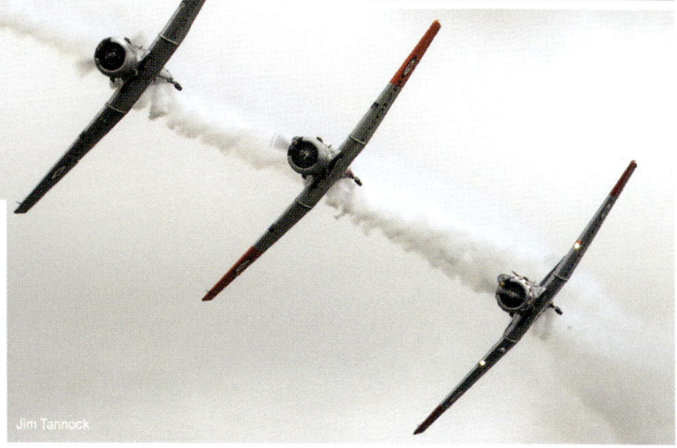
Jim Tannock

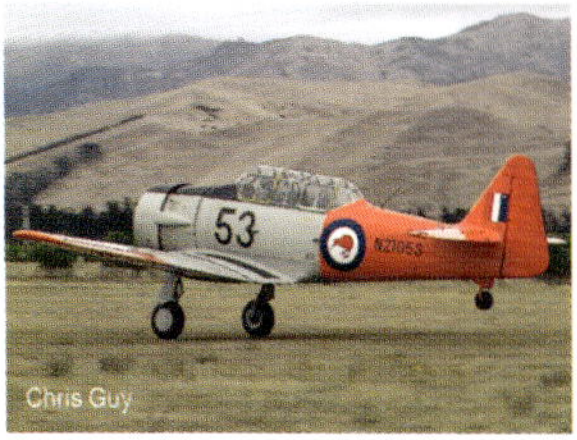

Chris Guy

Curtis P-40 Kittyhawks

Nothing gets an airshow crowd excited like the sound of a single-engined World War 2 fighter or two! As in 2003, this year we were privileged to see not one, but two P-40s at the show—the Auckland-based 'Australian' P-40N-1, and the 'American Volunteer Group (AVG)' P-40E now operated by the Old Stick and Rudder Company in Masterton. This latter aircraft is one of only two surviving and airworthy RNZAF Kittyhawks (the other is currently located in Australia).

This Page: The P-40's always seem to be in the thick of it, and always provide a great spectacle for the crowd. **Facing Page:** The Australian marked P-40N-1 ('Currawong') leads the OSRC's 'Flying Tigers' P-40E in a high speed dash along the crowd line. Photo Philip Merry.

North American P-51D Mustang

One thing that the Classic Fighters organisers enjoy and always strive to do is to bring you aircraft that you've never seen before—or at the very least, the same aircraft in new, different and interesting colour schemes. The French (i.e. European) theme for this year's show lead us to the idea of displaying an RAF P-51 this year, instead of the usual NZ Territorial Air Force one. So prior to the show, Graham Bethell's ZK-TAF was cunningly 'converted' into Flight Lt J. Dooley's aircraft 'Dooleybird', which flew with 19 Squadron RAF during 1945.

Erin Boyd

Gavin Hadfield

Top Left: Martin Nichol makes a start on painting the blue checkerboard pattern on the aircraft's nose. **This Page**: The Mustang wasn't used by the RNZAF during WW2, but 30 were operated in NZ between 1945 and 1957. These aircraft were the first (and ultimately only) batch of several hundred that were planned to replace the RNZAF's Corsairs in the Pacific Theatre. **Facing Page**: The Dooleybird makes its first appearance at this year's show, prompting many airshow veterans to wonder where this 'new' P-51 had come from. Photo Erin Boyd.

Alex Mitchell

Geoff Sloan

Gavin Conroy

Geoff Sloan

94 Classic Fighters Marlborough 2005

Goodyear FG-1D Corsair

This only surviving airworthy RNZAF Corsair returned to New Zealand in early 2004, and is now owned and operated by the Old Stick and Rudder Company in Masterton. Expertly handled by experienced Corsair pilot Keith Skilling, the crowd at Classic Fighters this year was witness to a superb display from this pilot and aircraft.

This Page: A striking fighter aircraft, in a beautiful display location. **Facing Page:** Kiwi pilot Keith Skilling provided the crowd with a great display of the aircraft's abilities. Photo Philip Merry.

Mike Hodgkinson

Philip Merry

Gavin Conroy

Alex Mitchell

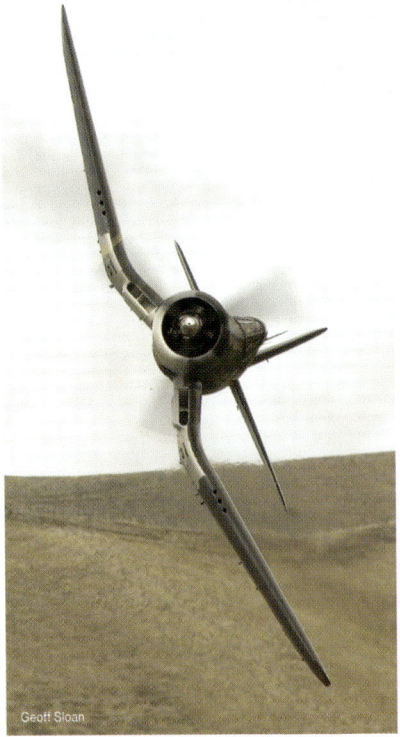
Geoff Sloan

WW2 Reenactment

Initially the World War Two scenarios at Classic Fighters 2005 were to cover several engagements in France during 1944—the attack on Carentan, the Failaise Gap, and a heroic battle by German Panther tank commander Ernst Barkmann. As mentioned previously by Dave Lochead (Ground Theatre Coordinator), with so many vehicles and reenactors to deal with things didn't go strictly to plan in this regard. But as the photos on the next few pages show, the WW2 performances were still explosive action-filled melées, almost worthy of a big budget Hollywood movie.

Geoff Sloan

Lyle Pethig
Gavin Hadfield

Erin Boyd

Geoff Sloan

This Page: The German forces strike out against the Allies.
Facing Page: It wasn't only the 'regular' troops who turned up for battle this year—we also had a few rather feisty French Resistance fighters—but unfortunately they got themselves captured. *Viva La France!* Photo Erin Boyd.

Jim Tannock

Alex Mitchell

98 Classic Fighters Marlborough 2005

Jim Tannock

Paul Maggs

Phil Teague

Geoff Sloan

This Page: The Allies strike back. **Above**: Field Marshal Bernard Montgomery (aka Chris Rhodes) is greeted by BSM Frank Walmsley before inspecting the Allied troops prior to battle. **Below**: Sometimes the fighting got so close, it was hand-to-hand combat that won the day. **Facing Page:** The Allied fighters scream in to provide close support. Photo Philip Merry.

Erin Boyd

Karen Mitchell

Gavin Hadfield

Karen Mitchell

100 Classic Fighters Marlborough 2005

Come and stay awhile...

The Marlborough Motel Association Inc

Quality motel and motor lodge accommodation guaranteeing good service is the hallmark of the Motel Association of New Zealand, the country's motel industry trade body with over 1000 members.

Marlborough has more than 30 MANZ motels and motor inns, all providing customers with courteous, friendly, prompt and honest service; clean, tidy and hygienic premises; and environmentally responsible operations.

Try the MANZ motel experience for a unique way to enjoy the beauty and hospitality of Marlborough. Local MANZ self-contained units are typically spacious, affordable, have modern kitchen facilities, easy parking and are located close to many local attractions.

Depending on location, you can choose from shoreline settings, vineyard views, mountain back-drops or town-centre conveniences. Accommodation styles include self-contained studio units, units with separate bedrooms, luxury apartments and honeymoon suites.

Marlborough Motel Association Inc.
PO Box 834, Blenheim, New Zealand
Email: enquiries@marlboroughmotels.co.nz
stay@brydan.co.nz

www.marlboroughmotels.co.nz

Paul Maggs

Karen Mitchell

Photographers & Reprints

A number of photographers have provided their images from Easter 2005, in order that we may bring you the many fine photos that grace the pages of this publication. Not only have we had the opportunity to print photos from the official team, but a number of other pro-am photographers have also kindly allowed us to reprint their images.

Like the many other volunteers who help make Classic Fighters the great show it is, these photographers have provided their time, and their lasting images, free of charge.

The photographers listed below are happy to sell you prints of any of their photos that appear in this book, or any of the literally hundreds of photos they took during the show. Please contact the appropriate photographers using their contact details below, and feel free to provide any feedback if you wish to do so. For anyone not listed here, please email:

    admin@classicfighters.co.nz

Alex & Karen Mitchell
    http://www.warbirdsovernewzealand.com

Philip Merry - pmerry@clear.net.nz

Geoff Sloan - geoffsloan@xtra.co.nz

Phil Teague - pmteague@xtra.co.nz

Jim Tannock
    http://www.jimtannock.co.nz

Gavin Conroy
    gavinconroy@xtra.co.nz

Paul Maggs - pmaggs@ihug.co.nz

Gavin Hadfield/Langwoods Photo Centre
    langwoodsblenheim@xtra.co.nz

Lawrence Ackett - llacket@xtra.co.nz

Erin Boyd - erin.e@xtra.co.nz

Chris Guy
    http://www.taupoairphotos.co.nz

D.L.A. Turner : 027 226 7345

Craig Justo - craig@aeroaspects.com.au

Rob Duff/Topshots Photolab
    http://www.topshots.co.nz

Erin Boyd By Gavin Hadfield

Phil Teague By Geoff Sloan

Website & Merchandise

Keep up with news and developments at Omaka, plus information about the next Classic Fighters airshow, by checking out our web site (see the address below). We keep the site updated on a regular basis, and you can also subscribe to our email newsletter so that you get notified of news and events via email almost immediately.

As we get closer to the next show, we usually run a competition or two so that you can try and win free tickets and/or other Classic Fighters items. Check out the web site and you can purchase a range of merchandise and memorabilia such as:

- Videos/DVD's
- Badges
- Caps
- Fleece Vests
- T-Shirts
- Models
- Glassware
- Posters, and more

Find us at: www.classicfighters.co.nz

AIRCRAFT INDEX

Aircraft	Page
Sopwith Camel	10
Fokker Dr.1 'Triplane'	12
Bristol Fighter F2.b	14
Halberstadt D.IV	16
Airco DH-2	18
Airco DH-5	20
Pfalz D.III	22
Nieuport 24	24
Reenactment	26
WW1 Reenactment	30
Pither Monoplane	34
Brough Superior Motocycle	36
Percival Proctor V	38
de Havilland Airliners	40
Beech Model 17 'Staggerwing'	42
Lockheed Electra 12a Junior	44
Fleet 16b Finch	46
Other Sights	48
Static Aircraft	52
Boeing 757 - RNZAF	54
CT-4 Airtrainer - RNZAF	56
Lockheed P3 Orion - RNZAF	58
Bell UH-1H Iroquois / Kiwi Blue	60
Restoration Row	62
Last Flight Of The Red Baron	64
Cri Cri	66
Percival 'Piston' Provost	68
Thunder Mustang	70
Yakovlev Yak-52	72
Nanchang CJ-6a	74
Soko Galeb G2-A	76
de Havilland DHC-1 Chipmunk	78
Air Tractor AT-402B	80
Aerobatic Displays	82
Saturday Evening Dinner	84
Consolidated PBY5a Catalina	86
Douglas DC3 Dakota	88
North American Harvard	90
Curtis P40 Kittyhawks	92
North American P51-D Mustang	94
Goodyear FG-1D Corsair	96
WW2 Reenactment	98
Photographers & Reprints	103

Erin Boyd

Sponsors

Thanks must go to all the major sponsors of Classic Fighters 2005—without them this great show wouldn't have been able to go ahead.

In particular we must mention Ross & Barbara at Lawson's Dry Hills Winery—their enthusiasm and sponsorship of the show was great, and was very much appreciated. Likewise the considerable support we have received from Wyatt and Wilson Print in Christchurch has been immensely valuable—thanks.

104   Classic Fighters Marlborough 2005